Published by The Diving Bookshop Press
St Ives, January 2006

Copyright © The Diving Bookshop Press 2006

All rights reserved. No part of this publication may be reproduced, stored in or introduced into any retrieval system, or transmitted, in any form, or by any means (mechanical, electrical, photocopying, recording or otherwise) without the prior written permission of the author. Any person who commits any unauthorized act in relation to this publication may be liable to criminal prosecution and civil claims for damages.

10 9 8 7 6 5 4 3 2 1 0

ISBN 978-1-291-74078-3

This book is sold subject to the condition that it shall not, by way of trade or otherwise, be lent, re-sold, hired out, or otherwise circulated without the publisher's prior written consent in any form of binding or cover other than that in which it is published and without a similar condition including this condition being imposed on the subsequent purchaser.

A HISTORY OF DIVING

FROM THE

EARLIEST TIMES TO THE PRESENT DATE,

COMPILED CHIEFLY FROM THE RESEARCHES OF

THE LATE

JOHN WILLIAM HEINKE, Assoc. Inst. C.E.,

WITH NOTES AND ADDITIONS BY

F. W. HEINKE AND W. G. DAVIS.

THIRD EDITION.

HEINKE AND DAVIS,

2, BRABANT COURT, PHILPOT LANE.

LONDON
WATERLOW AND SONS, PRINTERS, GREAT WINCHESTER STREET, E.C.

1871.

INTRODUCTION.

In bringing this pamphlet before the notice of Engineers, Ship Owners, Merchants, Contractors and others, the compilers trust that the information with regard to submarine operations generally, contained in the paper, will prove interesting and valuable to those whose profession requires them to avail themselves of the services of a diver. The main part of the work is the result of the researches of the late John William Heinke, Assoc. Inst. C.E.; the remainder is principally composed of newspaper extracts, and accounts of various remarkable submarine undertakings, together with such explanations as are necessary, from time to time, to render the knowledge of the use of the apparatus clearer.

At the end will be found a few plain practical directions for using Heinke and Davis's Diving Apparatus.

A HISTORY OF DIVING.

The following is a copy of a Paper "On Improvements in Diving Dresses and other Apparatus for Working under Water,"[1] by Mr. JOHN WILLIAM HEINKE, Assoc. Inst. C.E., which was read at a Meeting of the Institution of Civil Engineers, on March 18, 1856,

Mr. ISAMBARD K. BRUNEL, Vice-President,
being in the Chair.

"FROM the earliest times, divers have been in requisition for various purposes,—in raising goods from wrecks,—in recovering anchors, or lost treasure,—as well as in pearl, sponge, and other fishing,—and they, most probably, received no artificial aid; but practice, as in all pursuits, rendered some more daring, or greater adepts at the employment, than others, until, for the sake of ease and comfort, and greater security in following their avocation, different contrivances were resorted to, of which no accounts have been recorded. {Diving among the ancients.}

"Alexander the Great, at the siege of Tyre, employed divers to impede, or destroy, the works of the besieged, as they erected them. The Syracusans, also, trained persons for the same purpose, and for getting beneath and injuring the enemy's vessels. The Rhodians had a law, by which all divers were allowed a proportionate share of recovered treasure, according to the depth from which it was brought, or the risk incurred. If the divers raised it from the depth of eight cubits, or two fathoms, they received one-third;—if from 15 fathoms, one-half;—but for goods cast near the shore, and found at one cubit, or 18 inches from the surface, only one-tenth was paid. {Services of divers in time of war. Treasure Trove. Rate of compensation.}

"Among the pearl fishers, one method of prolonging operations {Pearl fishers.}

---
[1] Vide Minutes of Proceedings Inst. C.E., vol. xv., p. 309.

under water, was by retaining in the mouth, while below, a piece of sponge soaked in oil, and replenishing it at each plunge; this is said to have enabled some of the most hardy to remain at least six or seven minutes under water, which length of time is considered by physiologists to be the limit of submarine endurance.

Astracan divers. "The divers of Astracan used to step from a warm bath into the water, where they would continue six or seven minutes; they then returned to the warm bath, from which they again plunged into the water. This was repeated about five times during the day, after which they became exhausted and senseless, blood flowing from the nose and ears.

Dutch divers. "There are, however, many accounts of long continued immersion, which, if true, certainly show that other than natural means were resorted to for enabling men to work under water. The Dutch were originally celebrated as divers, and some are said to have remained below more than an hour. Mersennius, in his work, published at Amsterdam in 1671, mentions a man, named John Barrinas, who could remain six hours under water, which it is evident could only have been accomplished by the aid of mechanical apparatus. Many instances of intrepidity and dexterity in diving may be found in 'Jans Kraft, Sitten der Wilden, Kopenhagen, 1766;' but perhaps the most remarkable is that recorded by Father Kircher of the Sicilian *Nicolò Pesce.* diver Nicolò Pesce; his skill was amazing, and it was said that he carried letters for the King from Sicily to Calabria. His fame being wide spread, the King offered him a golden cup to explore the terrible gulf of Charybdis, and he remained three-quarters of an hour amidst the foaming abyss; on his return he described all the horrors of the place, and so astonished the monarch, that he requested him to dive once more, further to ascertain its form and contents. He hesitated, but upon the promise of a still larger cup and a purse of gold he was tempted to plunge again into the gulf, whence he never more emerged.

Necessity of artificial aid. "As, in the earlier stages of commerce, vessels, whether freighted or not, never ventured far from the shore, there were numerous opportunities for the diver, and as the chief considerations were time and safety, an apparatus that would supply fresh air for a comparatively long period was a great desideratum, for without such an instrument frequent visits must be made to the surface, causing an evident loss of time, and the mere act of holding the breath almost precluded the exertion of any physical force.

Invention of the diving bell. "Although the invention of the Diving Bell has been generally assigned to the sixteenth century, yet there are evidences of other, although rude, modes having been adopted long anterior to that era. Beckmann, in 1770, quoted a passage from Aristotle (Problem xxxii.) to show that the divers of his time used a sort of kettle to enable them to remain longer under water, but the inference drawn does not seem very clear. The renowned friar, Roger Bacon, who flourished about 1250, has been considered by some to be the originator of the diving bell, or of some machine to facilitate working under water, but little credit can be attached to the tradition. The earliest mention of anything of the kind that can be depended upon is by John Taisner, who says he saw the experiment made by two Greeks at Toledo, in Spain, in 1538, before the Emperor Charles V. and ten thousand spectators. Gaspar Schott (Nurnberg, 1664) repeats the story, and calls the vessel an 'aquatic kettle,' but prefers another apparatus, which he designates 'aquatic armour,' that enabled those covered with it to walk under water. The plate, accompanying the description, represents a man walking in the water with something like a small diving bell over his head. In an edition of Vegetius on 'The Art of War,' published in 1511, there is an engraving which represents a diver with a cap on his head, from which rises a long leather pipe, with an open end floating on the surface of the water. Vegetius.

"Lorini, in his work on Fortification, published at Venice in 1609, gives a plate and a description of a diving apparatus, or chest, which he described as 'a square box bound round with iron, furnished with windows, and has a stool affixed to it for the diver;' but he does not lay claim to the discovery, and seems to consider it as a machine already known. Lorini's apparatus.

"Francis Kessler, of Oppenheim, in 1617, gave a description of a suit of water armour, which, however, Beckmann[1] and others declared could not be used with safety.

"About 1620, Cornelius Debrell contrived a submarine vessel, or boat, to be rowed and used under water. This was tried upon the Thames by order of James I., and is said to have succeeded admirably; it carried twelve rowers besides passengers. This vessel is mentioned by Robert Boyle, in his 'New Experiments Physico-Mechanical,' wherein he professes to give Debrell's secret, from authority; he says :—'The composition was a liquid, that would speedily restore to the troubled air such a proportion of vital parts as would make it again, for a good while, fit for respiration.' This novelty induced 'His Most Serene Highness Charles, Landgrave of Hesse Cassel,' to have a diving vessel constructed for the same purpose: some Submarine boat.

---

[1] Vide Beckmann, translated by William Johnson, 8vo., London, 1797 and 1814.

years afterwards, it was described (with a diagram) in the 'Gentleman's Magazine.'[1]

Lord Bacon's apparatus.

"The celebrated Lord Bacon, in his 'Novum Organum,' 1645, suggested a machine, where the diver 'stood upon a stool of three feet as a tripod, which were in length somewhat less than a man, so that the diver, when no longer able to contain his breath, could put his head into the vessel, and having breathed again, returned to his work.'

Bishop Wilkins.

"Bishop Wilkins, in his 'Mathematical Magic,' 1648, proposes a machine 'whose benefits shall be incalculable: 1st.—Privacy, as a man may go to any part of the world invisible, without being discovered or prevented. 2ndly.—Safety from the uncertainty of tides and tempests, that vex the surface, from pirates, robbers, and ice, which so much endanger other voyages towards the poles. 3rdly.—It may be used to undermine and blow up a navy of enemies, or to relieve a blockaded place.' His plans were, however, wholly theoretical.

Original inventor of the torpedo.

"In 1663, the Marquis of Worcester published the heads of his 'Century of Inventions,' the necessary directions for carrying out these projects having been lost, as it is stated in his observations on the title-page. Proposal 9 is—'A ship-destroying engine, portable in one's pocket, which may be carried and fastened on the inside of the greatest ship, and at any appointed minute, though a week after, either of day or night, it shall irrecoverably sink that ship.' Proposal 10.—'Away from a mile off, to dive and fasten a like engine to any ship, so as it may punctually work the same effect, either for time or execution.' Proposal 11.—'How to prevent both.—How to prevent and safeguard any ship from such an attempt by day or night.'

Borelli's apparatus.

"In 1669, Borelli contrived a 'vesica' or bladder, which was, in fact, a copper vessel, two feet in diameter, with glass fixed before the face of the wearer. This contained the diver's head, and was fixed to a goat-skin habit, exactly fitted to the shape of the body. He carried an air-pump by his side, by means of which he condensed, or rarified the air in the vessel, and thus made himself heavier or lighter on the same principle as fishes. Within this 'vesica' there were pipes, by means of which a circulation of air was contrived; thus equipped, and with artificial webbing to the feet, to enable him to tread the water, the inventor supposed that he had overcome all difficulties hitherto known, or objections to which such machines were

---

[1] Vide Gentleman's Magazine, December, 1747.

liable. Hooke, in 1671, in his 'Philosophical Collections,'[1] speaks of Borelli's apparatus, of which he gives a plate, as also of one which he claims to have himself constructed. This he described as 'another way of swimming under water, and breathing, by the help of a leather pipe, kept open by wreathed wires, and extending from the diver's head to the top of the water.' In 1678, a German, named Sturm, was enabled to make some further improvements in Borelli's apparatus, but neither seem to have answered the intention, or ever to have been used.[2]

"Mersennius, in his publication at Amsterdam, in 1671, proposed a submarine boat, by which persons might pass from place to place under water, move it to and fro and make it rise or sink in a river or sea; this project failed, however, like all its predecessors. Another proposal for a diving machine appears, also, in the same place and in the same year, from Nicolas Witsen; he describes his invention very explicitly, and gives instructions to the divers as to its proper use and management, but there is not any account of its real utility or success. *[margin: Attempts to make submarine boats.]*

"About this time, a spirit of enterprise seems to have sprung up, and more attention was directed to the best means of securely searching for treasures hidden in wrecks, &c., and both in Holland and in Great Britain great efforts were made for that purpose. As might be expected, many of the schemes failed, either from want of proper machinery or lack of means; companies were formed in haste, for the purposes of exploration, and as quickly abandoned. The Duke of Argyll, among others, joined in the mania, and determined upon examining the wreck of a vessel sunk off the Isle of Mull, in 1588, being, in fact, one of the Spanish Armada, and supposed to be richly laden. He engaged for the task a man named Colquhon, of Glasgow, who went down several times, but merely surveyed the wreck as well as he could. The apparatus he employed seems to have been made after that suggested by Hooke, and consisted of a long pipe of leather, by which the air was communicated, his head being covered with some sort of bell. In 1688, Sinclair, a Professor in the University of Glasgow, published his Principles of Navigation, in which, in a postscript, he gives directions how to buoy up any ship of burthen, from the ground to the surface of the sea; and he speaks of the *[margin: Mania for treasure seeking. Duke of Argyll.]*

---

[1] Vide No. 2, p. 36. [2] Vide Collegium Curiosum.

Phipps, founder of the family of the Marquis of Normanby.

apparatus employed in searching the vessel, as being similar to that which Colquhon had previously used.

"The most successful adventure of the period was undertaken by one Phipps, a ship-carpenter, the son of a blacksmith at Boston, in America. He began to operate in 1687, with an apparatus, the character of which is now unknown, upon the wreck of a Spanish galleon, lying off the coast of Hispaniola; but what he then recovered did not repay the outlay. Nothing daunted, he determined upon trying again, and assisted with money (though most usuriously) by the Earl of Albemarle, son of the well-known General Monk, he eventually, but with much difficulty, rescued property of the value of nearly £300,000, of which sum he received about £20,000 for his own share. In other ventures he was equally successful; he was afterwards knighted, became Sheriff of New England, and founded the present noble family represented by the Marquis of Normanby.

"With the publication of a work by Pasch, at Leipzig, in 1700, the century closes. His plan was merely an alteration of others that had preceded, and was never, probably, tried with success.

Dr. Halley.

"The celebrated Dr. Edmund Halley, Secretary of the Royal Society, paid great attention to the subject for some years, and from his continued experiments, and the very different structure of his machine, he was considered as the inventor of the diving bell; that notion has, however, long been exploded.

"In 1716 he read his paper, entitled 'The Art of Living under Water,'[1] before the Royal Society; and the following extract conveys his views. He says:—'When there has been occasion to continue long at the bottom, some have contrived double flexible pipes to circulate air down into a cavity, enclosing the diver with armour, to bear off this pressure of the water, and give leave to his breast to dilate upon inspiration, the fresh air being forced down by one of the pipes, with bellows, or otherwise, and returning by the other of them, not unlike an artery, or vein. This has been found sufficient for small depths, not exceeding 12 or 15 feet, but when the depth surpasses 3 fathoms, experience teaches us that this method becomes impracticable; for though the pipes and the rest of the apparatus may be contrived to perform their office duly, yet the water, its weight being now become considerable, does so closely embrace and clasp the limbs that are bare, or covered with flexible covering, that it obstructs the circulation of the blood,

---

[1] *Vide* Phil. Trans., No. 349, vol. xxix., p. 492.

and presses with so much force on all the junctures, where the armour is made tight, with leather skins, or such like, that if there be the least defect in any of them, the whole engine will instantly fill with water, endangering the life of the man below. . . . . To remedy these inconveniences, the diving bell was next thought of.'

"He then describes his contrivance, which was a truncated cone of wood, containing 60 cubic feet in its concavity, the diameter at the top being 3 feet, and at the bottom 5 feet. In the top was placed a strong clear glass to give light, and a cock to let out the air that had been breathed. The machine was coated with lead and otherwise weighted, that it might sink steadily; when below it was supplied with air by two barrels of 36 gallons each, which were alternately lowered and raised, full and empty. In this instrument, he says that he remained without inconvenience, wholly dressed, with all his clothes on, for one hour and a half, at a depth of 10 fathoms. He subsequently conceived a method by which the diver could leave, and walk about some distance. This he also described to the Royal Society in 1721.[1] He says :—'I bethought myself how to enable the diver to go out of the bell to a considerable distance, and to stay a sufficient time without it, with full freedom to act as occasion served. . . . . I procured pipes to be made, which answered all that was hoped from them. They were secured against the pressure of the water by a spiral brass wire, which kept them open from end to end.' This appears to have been an adaptation of Hooke's apparatus, or of that used by Colquhon. These wires, of which the diameter of the cavity was about one-sixth of an inch, 'were coated with thin glove leather, curiously sewed on, and then we dipt the leather into a mixture of oil and bees-wax hot. Then we drew several folds of sheep's guts over them, which when dry we painted with a good coat of paint, and then secured the whole with another coat of leather, to keep them from fretting. The pipes, of which we made several, were about 40 feet long, the size of half-an-inch rope; the one end thereof being fixt in the bell, and the other fastened to a cock, which opened in the cap. . . . . The diver, therefore, putting on his cap, and coiling his pipe on his arm, like a rope, as soon as he is discharged from the bell, opens a cock and marches on the bottom of the sea, seeing that the coils of his pipe, which serves as a clew to direct him back again, &c., &c. . . . . The leaden caps were made

Crude diving bell.

---

[1] *Vide* Phil. Trans., No. 368, vol. xxxi., p. 177.

to weigh half a hundred weight, to which I added a girdle of large weights of leads, of about the same weight, in the whole, to be worn about the waist, and two clogs of lead for the feet, of about 12lbs. each.'[1]

Lethbridge's apparatus.

"About the same time that Dr. Halley read his first paper to the Royal Society, in 1716, a person named John Lethbridge, of Newton Abbot, near Exeter, invented a machine, which was made under his directions by a cooper in Stanhope Street, Clare Market, the particulars of which he published about thirty-three years afterwards in the 'Gentleman's Magazine.'[2] He thus describes it:—'It is made of wainscot, perfectly round, about 6 feet in length, about $2\frac{1}{2}$ feet in diameter at the head, and about 18 inches diameter at the foot, and contains about 30 gallons. It is hooped with iron hoops without and within, to guard against pressure; there are two holes for the arms, and a glass about 4 inches diameter and $1\frac{1}{4}$ inch thick, to look through, which is fixed in the bottom part, so as to be in a direct line with the eye; there are two air-holes upon the upper part, into one of which air is conveyed by a pair of bellows, both of which are stopped with plugs immediately before going down to the bottom. At the foot, there is a hole to let out water; sometimes there is a large rope fixed to the back, or upper part, by which it is let down, and there is a line called the signal-line, by which the people above are directed what to do, and under is fixed a piece of timber as a guard for the glass. I go in with my feet foremost, and when my arms are got through the holes, then the head is put on, which is fastened with screws. It requires 5 hundred weight to sink it, and taking 15 lbs. from it, it will buoy upon the surface of the water. I lie straight upon my breast, all the time I am in the engine, which hath many times been more than six hours, being frequently refreshed upon the surface, by a pair of bellows. I can move it, about 12 feet square at the bottom, where I have stayed many times 34 minutes. I have been 10 fathoms deep many hundred times, and have been 12 fathoms, but with great difficulty.'

Symond's machine.

"Another claimant appeared, nearly at the same time, in the person of Nathaniel Symonds, of Harburton, near Totnes. He produced a diving machine in the shape of a boat, in which, before many hundreds of persons, he sank himself in the River Dart, where he remained three-quarters of an hour, and then reappeared. He complains, with evident disappointment, that

---

[1] *Vide* Plate 2 in vol. iii. of Plates, Rees' Cyclopædia.
[2] *Vide* Gentleman's Magazine, Oct., 1749.

'though a great number of gentlemen of worth were present, he received but one crown piece from them all.'[1]

"In 1724, Jacob Leupold, of Leipzig,[2] describes an apparatus, then in vogue, but of which he does not claim the invention. Later still, Martin Triewald, Military Architect to Frederick, King of Sweden, greatly improved upon Halley's invention by making a machine both lighter and less expensive. In the head of his apparatus, which was of tinned copper, and which was managed by two men, he used in lieu of plain glass, convex lenses to admit the light. He published the particulars at Stockholm, in 1732, and the description was subsequently read before the Royal Society.[3] *Martin Triewald's apparatus.*

"About 1750, a Mr. Rowe invented a 'diving engine' for searching wrecks, which consisted of a hollow copper vessel, of sufficient dimensions to contain the body of a man, with holes at the sides, through which his arms protruded. At the end of the 'engine' glasses were placed, through which he could see the objects of his search. The diver was lowered by a rope, and could remain below, for half an hour, without any difficulty.[4] *Rowe's diving engine.*

"A daring, but unfortunate, attempt to use a submarine vessel was made in 1774, by Mr. Day, in Plymouth Sound. So confident was he of success, that he had a small ordinary vessel prepared for the purpose according to his directions, and at the time appointed for making the experiment, all being ready, he sank the vessel and himself, in the presence of a great many spectators, but he never rose again. *Mr. Day.*

"In 1775, Mr. Spalding brought out an improvement upon both Dr. Halley's and Triewald's apparatus, and was rewarded by the Society of Arts[5] with twenty guineas; he was followed by Farey, who rendered it still more complete and more applicable to the required purposes. *Spalding. Farey.*

"About the same time (1775), a Mr. Bushnell, of Connecticut, endeavoured to realize the project of Bishop Wilkins, and with the apparatus he constructed, he offered the newly-formed Republican Government of America, to destroy the British shipping, then lying in their different harbours and rivers; but although he found it quite practicable to travel under water, he *Attempt by Bushnell, in America.*

---

[1] *Vide* Gentleman's Magazine, July, 1749.
[2] *Vide* Theatrum Machinarum Hydraulicarum, Leipzig, 1724-5.
[3] *Vide* Phil. Trans., vol. xxxix., p. 377.
[4] *Vide* Universal Magazine, Sept., 1753.
[5] *Vide* Transactions of Society instituted at London for the Encouragement of Arts, &c., vol. i., p. 220.

did not succeed in the rest of the design. His machine had a resemblance to two upper tortoise shells, of equal size, joined together, and it was capable of holding the operator, with sufficient air to support him for thirty minutes. He could swim so close to the surface of the water, as to approach, unperceived, very near any ship during the night. He could sink quickly, keep at any depth he pleased, could rise to the surface for fresh air, and descend again at pleasure, as described in his publication of 1787.[1]

Carried out by Colonel Colt.
"This scheme was resuscitated in 1822, by Mr. Samuel Colt,[2] who proposed to the Government of the United States, to construct a machine, which would effectually realize all that Bushnell had suggested. In order to test its utility, the Secretary of the Navy was instructed to render Mr. Colt every assistance and facility, and to appropriate 15,000 dollars for the purpose.[3] A vessel was actually destroyed at some distance from the shore, but the means employed were not made public.

Martin's apparatus.
"Benjamin Martin, originally a plough-boy in Surrey, but afterwards a celebrated optician and globe manufacturer in Fleet Street, published in 1778 a description of his diving apparatus.[4] It consisted of strong leather, so prepared that no air could pass through: it fitted his arms and legs, and had a glass window in the front part. This apparatus held half a hogshead of air, and when dressed in it he could walk on the ground, at the bottom of the sea, or enter the cabin of a submerged ship and take out any valuables. He appears to have used this apparatus rather successfully. In his work, he speaks of a machine for the same purpose, by a gentleman of Devonshire; it is presumed that he alludes to Lethbridge's apparatus, previously described.

Smeaton.
"Smeaton, in 1779, first employed the diving bell for Civil Engineering operations; it was used in repairing the Bridge at Hexham, in Northumberland.[5] The apparatus was an oblong wooden box, 4 feet high, 2 feet wide, and 3 feet 6 inches long. It was supplied with air by a pump fixed on the top. He afterwards constructed an improved apparatus, of which he made use in the construction of Ramsgate Harbour, in 1788. Mr. Rennie subsequently made great improvements in it, adapted it to local circumstances, and extensively used it at the works of Howth Harbour, near Dublin.

Hexham Bridge.

Rennie's improvements

---

[1] *Vide* Brewster's Edinburgh Cyclopædia, vol. viii., art. Diving.
[2] Now Colonel Samuel Colt, Assoc. Inst. C.E.
[3] *Vide* Nautical Magazine, vol. viii., New Series, 1844, p. 74.
[4] *Vide* Philosophia Britannica, 1778.
[5] *Vide* Smeaton's Life.

"The apparatus designed by Mr. Kleingert, of Breslau, was first described in a pamphlet published in 1798. The harness, or armour, was made of strong tin plate, in the form of a cylinder, which enclosed the diver's head and body; it consisted of two parts, that he might easily get it on. Besides this, he had a jacket with short sleeves, and a pair of drawers of strong leather, all water-tight, and joined by brass hoops round the metal on the outside, so that he was relieved from pressure on all parts, except the legs and arms. With this apparatus, on the 24th of June, 1798, a man named Joachim, under his direction and before many spectators, dived and sawed through the trunk of a large tree at the bottom of the River Oder, near Breslau. *[margin: Operations on the Oder.]*

"At this time there were many projects of analogous character, but none particularly worth notice, except that by Robert Fulton, who first introduced steam navigation on the rivers of America. At the close of the last century he made a submarine boat, or chest, which he exhibited, under the patronage and at the expense of the French Government, on the Seine at Havre and Rouen; and afterwards at New York, and other places in America. *[margin: Fulton's apparatus.]*

"In the year 1786, Messrs. John and William Braithwaite were engaged in recovering the guns from the floating batteries which were sunk off Gibraltar, and they presented eight pieces of fine Spanish ordnance to the Emperor of Morocco. In the years 1789 and 1790, they successfully searched for and recovered all the dollars, and a large quantity of tin and lead, from on board the 'Hartwell,' East Indiaman, lost off Bonavista, Cape de Verde Islands. This was accomplished in depths varying from five to seven fathoms, by means of Mr. John Braithwaite's diving machine. On their return from Bonavista they negociated with the Government to commence operations on the 'Royal George,' and they made all the necessary preparations. But although the ship ostensibly belonged to the Admiralty, its guns were claimed by the Ordnance: hence difficulties arose between the Government Departments, which induced Messrs. Braithwaite to relinquish their design. From the wreck of the ship 'Earl of Abergavenny,' outward-bound East Indiaman, of 1,300 tons burthen, they succeeded in recovering nearly all the cargo, and £75,000 in dollars. This vessel was lost in 1805, and after having laid under 10 fathoms water for 16 months, during which time many unsuccessful experiments were made by Mr. Tucker, Mr. John Braithwaite, by means of his peculiar diving machine, (but which was not a diving bell,) succeeded in raising the ship and cargo, amounting in value to many thousands *[margin: J. and W. Braithwaite. Recovery of treasure. Official difficulties. £75,000 recovered.]*

of pounds. By this apparatus he was enabled to remain under water eight or ten hours at a time, and to conduct the various operations, which were effected by machinery exclusively his own, and by the aid of gunpowder. The diving machine which he employed is now the property of his son, Mr. John Braithwaite (M. Inst. C.E.), who, with his brother, was present on several occasions to witness the operations.

Divers, equipped in Heinke and Davis's Improved Patent Diving Apparatus, building Submarine Foundation.

Plymouth breakwater. "The Plymouth Breakwater, for which many plans had been suggested, was commenced on the 12th of August, 1812, the first stones being deposited with much ceremony. In the progress of a work of such magnitude, extending over many years, and which, in fact, is scarcely yet completed, this mode of working below has been found of essential service, and has been adopted, wherever it has been necessary to have firm and substantial masonry constructed under water. At the works for the Harbours of Refuge at Dover and at Alderney, it is extensively used by Mr. Walker.

Dover and Alderney harbours.

Royal George. "But, perhaps, one of the most striking uses to which it has ever been applied, was in the demolition of the wreck of the ill-fated 'Royal George,' sunk off Spithead, in August, 1782. In less than a month after the accident, several proposals were made for weighing her, and the proposition of Mr. Tracey was selected from those of 118 candidates. It was no easy undertaking, considering that the weight of the guns, stores, &c., on board, amounted to 1,031 tons, and that she had sunk 13 feet into a bed of silt or blue clay. After a trial of three seasons, and an expense of £12,000, borne between the Government and Mr. Tracey, the project was abandoned. Thus matters rested until June, 1817, when Mr. Ancell, of the Portsmouth Dockyard, went down in a diving bell and surveyed the wreck, as far as was practicable, in a depth of water of ten fathoms. Another respite of 17 years then took place. In the meantime Mr. Deane, with an apparatus which was originally intended for the recovery of property from houses or factories while on fire, but which, having failed to obtain the patronage of the insurance offices, he applied to diving purposes, succeeded, in 1828, in clearing the wreck of the 'Carnbrae Castle,' Indiaman, lost at the back of the Isle of Wight; he also operated upon the wreck of H.M.S. 'Boyne,' burnt at the latter part of the last century off South Sea Castle. He then offered to remove the wreck of the 'Royal George,' and after some delay received permission to make the attempt. In 1834-5-6, having had more perfect apparatus made under his directions by Siebe, he was enabled to bring up 28 guns (of which 21 were of brass and in good preservation) and also some other portions of the wreck.

"His task being so far complete he attempted, with success, to bring up the guns of the 'Mary Rose,' which was lying not far from the other wreck. This vessel had been submerged for nearly 300 years, as she went down in July, 1545, and its situation was only discovered in 1836, by some fishermen, whose nets had sustained injury from something protruding from the bottom of the sea. In 1836 he succeeded in raising 25 guns, five of which were of brass, and the other 20 of wrought-iron. The brass guns bore date 1535, and the makers' names were 'Robert and John Owyn.' The other guns were of peculiar construction, being manufactured of wrought-iron bars, secured by 33 hoops; besides these, he brought up some iron and many granite shot, eight ancient bows, a number of miscellaneous articles, and part of the oak mainmast; the latter still in good preservation. *Deane's operations.*

"In 1839 the operations upon the 'Royal George' were resumed, under the direction of Colonel (now Lieutenant-General Sir C. W.) Pasley. It was determined, if possible, to clear the roadstead, although at one time doubts were entertained of the propriety of attempting it, unless it could be done so effectually as to leave the bank entirely free of all debris. The *"Royal George"—further operations.*

destruction of the remains by gunpowder having been resolved upon, cylinders were prepared, and being heavily charged, the first explosion was reserved for, and took place on, the 29th of August, 1839, that being the fifty-seventh anniversary of the melancholy event. On the 20th September a charge of 260 lbs. of powder was fired by the voltaic battery, and this is supposed to have been the first public practical adaptation of such means, although the applicability of the voltaic battery for such purposes had been previously demonstrated by Mr. J. Bethell (Assoc. Inst. C.E.) on the 24th of April, 1838, at the Institution of Civil Engineers. The effect was instantaneous, highly satisfactory, and grand beyond description. The surface of the

Great success. water was immediately covered with dead fish and with fragments of all descriptions, curiously covered with seaweeds, and of richly tinted colours. During the season they recovered, among other things, five brass guns weighing 26,072 lbs. (the value of which, as old metal, was estimated at £1,000), and seven iron guns.

Hall, of Whitstable. "In 1840 the Colonel again resumed operations and with the same success, having an able assistant as a diver in Mr. George Hall, of Whitstable, with about 80 men, including Sappers and Miners. The apparatus employed was manufactured by Siebe, by whom several alterations, at Deane's suggestion, had been made in that previously used.

Timber saved. "By the end of 1841 much valuable timber was rescued; indeed, between the months of May and November in that year, not less than 18,600 feet, or 372 loads, of timber were brought up, and being afterwards sold by public auction, great quantities were preserved as relics.

Weight of gunpowder used. "The season of 1842 was quite as satisfactory as those preceding, and in that of 1843 the harbour was finally cleared of all obstruction. The consumption of gunpowder during the operations was 52,963 lbs., and there were recovered no less than 581 cwt. 2 qrs. 14 lbs. of various sorts of metal (exclusive of 86 guns) and 59,000 cubic feet of timber.

Curious facts. 'It may be remarked, as somewhat curious, that of all the money which must have been on board at the time of the catastrophe, when 1,200 persons went down, only two guineas were found. It is, moreover, satisfactory to know that during the works no accident occurred attended with loss of life or limb, although there were three or four narrow escapes. But, perhaps, the most singular incident is that an actual fight took place below between two divers for the possession of some portion of the wreck claimed by both; in the scuffle the glasses of one helmet were broken and the diver was nearly drowned before he could be rescued. Such an accident is now, in Heinke's apparatus, effectually provided against by helmet slides.

"The operations against the wreck in question also resulted in great benefit to the practice of diving, the applicability of the apparatus having been tested in every possible manner. Many of the Sappers and Miners, both of the regular army and of the East India Company, were fully initiated in the use of diving apparatus, and sailors from different vessels were also trained so as to be useful in cases of emergency. *Impetus given to diving.*

"In France there has been many efforts towards establishing the practicability of submarine boats, but the great difficulty was how to supply air to the men employed. Dr. Payerne,[1] however, felt convinced that it was practicable, by chemical means, to restore the purity of the air under water without communication with the atmosphere. This experiment was first tried in England, at the Polytechnic Institution, and was then repeated with success at Spithead. *Dr. Payerne's scheme.*

"On the latter occasion, the bell was accompanied by four cylinders, each 4 feet long and 12 inches in diameter, containing condensed air, which was forced into them by an air-pump, and allowed, when required, to flow into the bell, by turning a cock. Another experiment was made without cylinders; the end of one of the diver's air-pipes was conducted into the bell, and air was forced through it by one of the small pumps ordinarily used for supplying air to a helmet diver. The water was kept out of the bell as well as under the ordinary system, and the respired air was renewed in a perfectly satisfactory manner. The result was approved by Lieutenant-General Sir C. W. Pasley and other scientific men, the air for respiration being perfectly good, and the whole apparatus for purifying it so compact and simple, that it could be contained in a case not larger than a common portable desk, and it could be used without any trouble.

"The helmet diving apparatus has now become of comparatively common use, for the repair of lock-gates and other works under water. In the building of almost all docks, bridges, &c., of any extent, it is in constant use; and in examining accidents occurring to vessels, and more especially to the shafts of screw-propellers, to rudders, &c., it has been very useful. It has been so constantly exhibited at the Polytechnic Institution and at the Panopticon, that it has become familiar to all. *Uses of the diving apparatus.*

---

[1] *Vide* "Description of a Diving Machine, employed in the Government Works of Cherbourg by Dr. Payerne:" by Captain H. Tyler, R.E. Published in the Part Papers of the Corps of Royal Engineers, vol. v., New Series, p. 35.

"Besides the various alterations and improvements already mentioned, there have been many others that deserve notice.

Bethell's apparatus.
"In 1835, Mr. J. Bethell introduced several important improvements in the form and use of diving apparatus.

Bush's improvements.
"In 1836, Mr. William Bush, of Bishopsgate, claimed the introduction of air-pumps into diving bells, instead of pumping air down from above; the application of a pump to diving dresses, whereby the diver might supply himself with air from above; and the use of an air-belt, combined with a diving dress, to facilitate the diver in rising and sinking. He also applied a compass to the helmet of the dress, in order that the diver might ascertain his position when below the water.

Frazer's improvements
"In 1836, Mr. Frazer made an improved escape valve, and other additions to the dress.

Thornthwaite's improvements
"In 1838, Mr. Thornthwaite, of Hoxton, produced a diving-belt, for which he was rewarded with a silver medal by the Society of Arts, and his invention was ordered to be placed in their Repository. His instrument consisted of a belt of India-rubber cloth, to which was attached a small strong copper vessel; into this, air was forced by a condensing pump, until it had a pressure of between thirty and forty atmospheres. The belt, being put on in a collapsed state, did not give any buoyancy, nor impede the diver in his descent. If he desired to rise, he opened a valve, by which the condensed air escaped from the metal vessel into the belt, and by its expansion, enabled him to rise to the surface.'

"In the ordinary apparatus, and hel-

FIG. 4.

Heinke & Davis's Helmet.

* Vide Transactions of Society instituted at London for the Encouragement Arts, &c., 1838-9-40, vol. lii., p. 243.

met, great alterations have been effected from time to time by the various makers. The improvement introduced by Mr. E. Heinke and the Author, some of which are shown in Figs. 1, 2, and 3, are based upon long experience of the defects complained of in ordinary apparatus. The submarine dress, as manufactured at that period, was exhibited at the Great Exhibition, in Hyde Park, in 1851, and obtained the award of a medal; since that time, several additional improvements have been made, and so nearly perfect may the apparatus be now considered, that there are few persons who would hesitate to go down in the dress, after once seeing it used.

Heinke's improvements.

Prize Medal obtained.

Heinke's valve.

Side view and sections of Heinke & Davis's Helmet.

"Among the most prominent of the improvements is that of a double valve fixed in the front of the gorget, which enables the diver to descend and rise at pleasure, with the whole of his gear, which weighs upwards of 200 lbs.; in fact, it places the whole apparatus completely under his control, and protects him in case of anything happening to the air-hose, as by its means a sufficient quantity of air to support respiration for ten minutes can be contained in the helmet and dress, thus giving time to ascend, even from a very great depth. The connecting joints are now so manufactured that they can scarcely be broken, as they will resist the most powerful pressure in consequence of having a double safety-cap affixed. The new vulcanized band completely excludes the water from the dress, and enables it to fit more easily and with greater comfort to the

Double safety cap joints.

wearer. The signal dial makes the wants of the diver known to those above, instantly and correctly, and, in fact, renders the apparatus nearly complete for the most difficult undertakings.

Experiments with Heinke's apparatus.

"In 1855, a number of interesting trials took place in various places. The experiments which were conducted at Portsmouth, in the month of June, in the presence of the Admiral-Superintendent and Dockyard Officers, gave great satisfaction, the diver remaining below half an hour at a time, in a depth of water of $3\frac{1}{2}$ fathoms. At Chatham Dockyard, in October, a similar trial took place, in the presence of the Captain-Superintendent and several gentlemen connected with the establishment, as well as many officers of the corps of Royal Engineers and others, who all expressed their gratification at the result of the experiment.

Trial competition at Paris, 1855. Heinke's declared by International Jury superior to Siebe's and Tyler's.

"At Paris, it was tested on the Seine, by command of the French Government, and in the presence of Prince Napoleon, a large number of military and other engineers, the Commander-in-Chief, the Secretary of the Exhibition, and the members of the International Jury.[1] On that occasion, five kinds of diving dresses were tried (of which, three were English and two were French),[2] in every variety of situation. The apparatus which was attended with the most successful results, and which, it was decided, possessed the greatest facilities, was that exhibited by Mr. Heinke. The diver, without any assistance, raised himself to the surface by partly closing the valve in the breast-plate of the helmet; the compressed air thus filled the waterproof dress, and brought him up. When he wished to descend, he had only to turn on the air-valve.

Quickness of Heinke's diver.

Severe test of Heinke's apparatus.

"In order to test the alertness of the divers, twelve small rings were thrown into the river; of these, Heinke's diver picked up ten, and the other two were not found by any of the divers. Again, at the request of Prince Napoleon, he went down with a helmet the glass of which had been accidentally broken at the Westminster Bridge works, and which, of course, admitted water; he immediately closed the safety-valve as directed, and remained under water half an hour before coming to the surface. The underclothing was then examined, and was found to be perfectly dry. The divers representing the other makers were then requested to submit their apparatus to the same test, but they all declined. The French exhibitor, M. Ernoux, at once, and in the most handsome manner, acknowledged the

Other makers requested to submit to same test. They decline.

---

[1] *Vide* Morning Post, Oct. 1st, 1855.
[2] The English makers were Siebe, Tyler, and Heinke. French: Ernoux and Cabirol.

superiority of Messrs. Heinke's apparatus, stating that he considered it really perfect.

"At the close of the Paris Industrial Exhibition, a First Class Medal was awarded for the apparatus, which was transferred to the Crystal Palace, at Sydenham."

THE apparatus will be found, on experience, to be the most convenient and comfortable of any manufactured. The valve can be adjusted by the diver without coming to the surface of the water, and regulated so as to supply air of any given density, without alteration of the rate of pumping. Heinke and Davis's apparatus.

It has been found that the more simple way for the diver to communicate with those above is by means of the life line (a strong rope fastened round the diver's waist and held at the surface by the signal man), one pull indicating "all right," two pulls "come up," and so on. The code of signals is not by any means arbitrary, nearly every clique of divers using dissimilar ones. Some, by means of long experience, have got so expert that they can hold a conversation by means of pulling at the life line. Signalling.

The dress is made of the best vulcanized India rubber, covered with stout twill inside and out. The material is very strong, but yet not too thick to obstruct the diver in his movements. The hose is also of best vulcanized India rubber, with a spiral wire in the interior to keep it always open, and afford a free passage for the air. Dress.

The helmet is made of tinned copper, with brass eye frames, neck rings, &c. It has an arrangement for distributing the air equally all over the helmet. It is fitted with Heinke's valve, which, besides the advantage of adjusting the pressure of the air, also enables the diver to rise to the surface, or sink to the bottom of the water, at pleasure. Helmet.

The boots, weights, under-clothing, knife, belt, &c., will all be found of the most substantial kind. The total weight borne by the diver under water, as supplied in the apparatus of Heinke and Davis, is about 200 lbs. including helmet and dress. Complement of the apparatus.

24

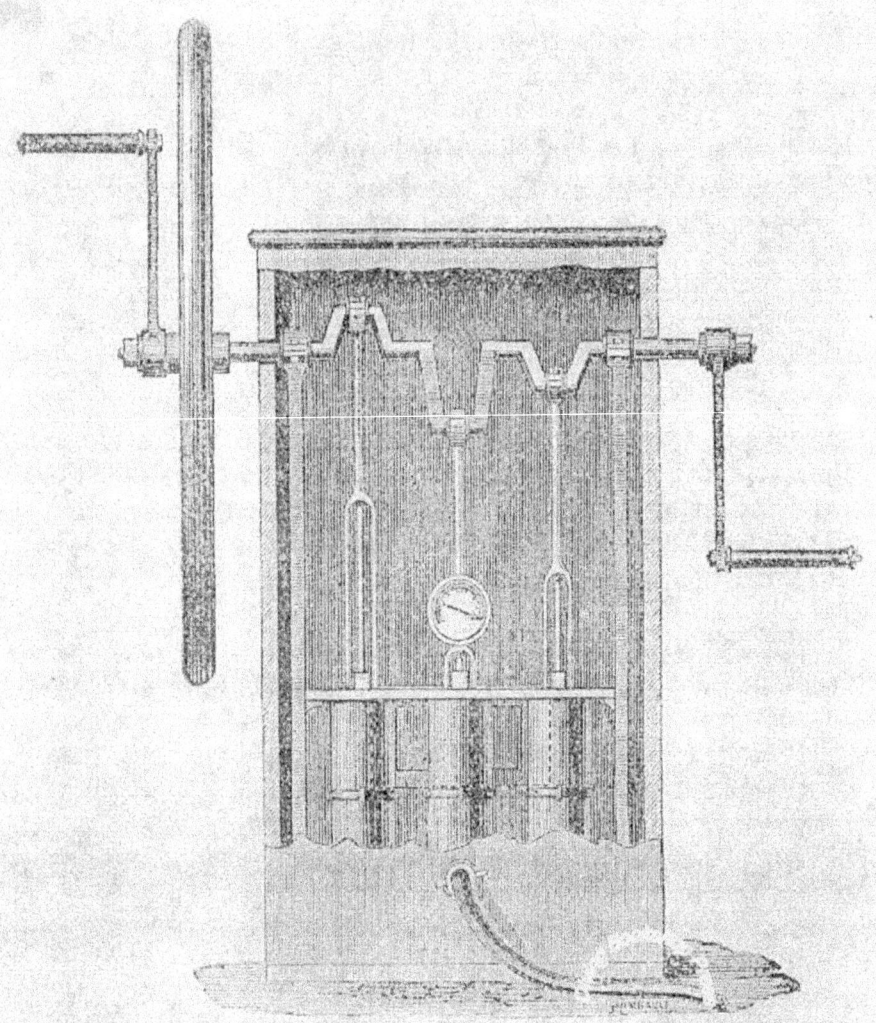

Heinke & Davis's Improved Patent Air Pump and Fire Engine combined.

Improved Air Pump and Fire Engine.

Heinke and Davis specially recommend to the notice of shipowners their **Improved Air Pump and Fire Engine combined.** This improvement not only supplies air to the diver when under water, as in the ordinary pump, but can also be used as a fire engine. It will deliver water to a considerable height, and would prove invaluable in case of any of the magazines of a ship catching fire. It may be kept ready for use night and day without interfering with the ordinary business on board ship. It can also be used for cleaning the decks, wetting the sails when the wind is slack, and in case of the engine pump (supposing the vessel to be a steam ship) breaking down it may be used as an auxiliary pump to fill the boiler.

Necessity of having a diving appa-

This, in conjunction with the diving apparatus, should form part of the equipment of every vessel, whether of war or peace, as in case of the paddle, screw propeller, or rudder getting out of order, in the event of a leak, which would usually be fatal, ratus on all vessels. or should the ship's bottom become covered with weeds or barnacles, the assistance of the diver is invaluable, and many good ships and human lives might have been saved from a watery grave had there been a diver and apparatus on board to survey and stop the leak. There would then be no heartrending tales of men working at the pumps till they could scarcely lift their weary arms; feeling the water gaining on them inch by inch, yet not daring to rest even for a moment, and at last, when hope of gaining on the leak was no longer possible, having to take to sea in small open boats, to traverse the ocean and to perish by hunger and thirst, unless picked up by some vessel.

The cost of the complete apparatus is comparatively small, and sailors, or others, are instructed by Heinke and Davis in the art of diving.

Heinke and Davis have also introduced another improvement into their pump, which much tends to the economical working of the apparatus. In the pump as usually supplied, one man only can go down at a time. This necessitates two pumps, and, as a matter of course, two men to work them. The wages of these men average £1 per week each, so that by means of Heinke and Davis's improvement a net saving is effected as follows:— Cost of extra pump and £100 per annum in wages. The improvement is simple and effective, and by its aid one or two men, as may be desired, can be sent down at once from one pump.

Improvement for sending down two men at once.

Great economy in working.

Divers equipped in Heinke and Davis's Improved Diving Apparatus. Freeing Screw Propeller from Weeds, and Repairing Leak.

Accident to H.M.S. "Thunder," repaired by diver.

As an instance of the use and true economy of the diving apparatus, the following facts are appended:—In January, 1837, H.M.S. "Thunder," striking upon a rock in the vicinity, sustained such severe injury, that she was run upon a sand-bank to prevent her foundering. It was almost decided to land her stores at Nassau, and break her up; but finding a diving apparatus at hand, the carpenter, by its aid, was enabled in twenty minutes to repair the damage. The ship was got afloat again, and on its return to Portsmouth, it was admitted that *the repairs could not have been better executed in the dockyard*. As all are unfortunately aware, it is most difficult to move the Government authorities upon any scientific subject: but, after this successful trial, the Admiralty ordered a few dresses, one of which was supplied to H.M.S. "Wellesley."

Accident to H.M.S. "Wellesley."

In 1838, that vessel, in entering the harbour of Trincomalee, met with an accident similar to that which occurred, the year before, to the "Thunder." She was on the point of sinking, for she made water so fast, that it could not be kept under with the pumps, when the diving apparatus was rigged, and although the ship was in 20 fathoms water, twenty minutes were sufficient to stop the leak; she afterwards proceeded on her voyage to China, and is, it is believed, still afloat.

Saving anchor or chain cable.

In case of an anchor fouling or breaking from the cable, or chain cable being broken, the diver can always recover it, and thus prevent a considerable loss.

Use to gas and water companies.

To gas and water companies, Heinke and Davis's apparatus will be found most useful. To the gas companies for examining the bottom and interior of gasometers; to water companies for examining and repairing valves, &c., without drawing off the water. As an instance of the success with which it may be applied to gas works, the following is quoted verbatim from the Nottingham *Daily Guardian*, August 11th, 1865, premising however, that the diver, G. Smith, referred to is on the staff of Heinke and Davis:—

1.—DIVING EXTRAORDINARY.

"During the past fortnight the Nottingham Gas Company
"have been engaged in erecting a new gas holder in the Poplar
"Works, at the end of Island Street, London Road. Shortly
"after the undertaking, the resident engineer (Mr. Lomas)
"ascertained that the clay and puddle at the bottom of the tank
"was unequal, and the result was that an application had to be

"made to Mr. G. Smith, the experienced diver, who recovered
"the watches and jewellery which had been stolen from the
"establishment of Messrs. Walker, Cornhill, London. Mr. Smith
"recovered ten watches from the bed of the Thames, the value of
"one of which was upwards of £50. He has also been successful
"in recovering valuable property in other places, and his opera-
"tions in Nottingham have excited much curiosity. The appa-
"ratus and dress used by Mr. Smith are the invention of Mr. W.
"Heinke, A.I.C.E., of Great Portland Street, London, and are
"of a very ingenious description. They enable Mr. Smith, with
"the greatest safety, to perform his arduous duties under water of
"the most disagreeable nature. A gas tank is often of the most
"repulsive nature to the eye of ordinary mortals, but Mr. Smith
"has, nevertheless, been able to remain as long as five hours in
"one at the Nottingham Gas Works without coming to the sur-
"face. While he has been engaged at the works a large number
"of persons have visited the place and inspected his novel diving
"apparatus and dress (which is made of vulcanized India rubber)
"with much interest. We may state that on reaching the sur-
"face, by means of a ladder, he took a leap and at once dis-
"appeared, having several weights attached to enable him to
"sink. When at the bottom of the tank, he was in utter dark-
"ness, but was nevertheless able to satisfactorily carry out the
"work entrusted to him. During his operations, every communi-
"cation was transmitted to him by his son (who remained on the
"bank) through the signal line, and he was supplied with air by
"a rotary three-throw pump. In addition to making the bed of
"puddle all right, Mr. Smith, during the past week, placed 16
"iron plates upon which the gas holder rests at the bottom of the
"tank. In this undertaking he has been entirely successful,
"and afforded the highest satisfaction both to Mr. Hawksley,
"the chief, and Mr. Lomas, the resident engineer."

It will be seen from the above that the use of Heinke and Davis's apparatus is by no means limited, but that it can be used for all purposes where water, foul air, or gas preclude ordinary labour. Heinke and Davis strongly recommend a periodical inspection of the parts of bridges, harbours, sea walls, pile foundations, ships' bottoms, &c., which are below water. A small flaw or defect by the action of the tide rapidly becomes a large one, and as it is utterly impossible to detect the faults without the aid of a diver, the damage is not perceived until it declares itself in the most emphatic manner by a wall caving in or a pile breaking, and in fact causing a loss of perhaps many thousands of pounds, besides the delay and feeling of insecurity always engendered by such an accident. The timely prevention

Application of apparatus to many purposes.

Inspection of submarine works.

of this by means of a diver will save thirty or forty times the cost of the apparatus.

Phosphate of Lime. — Heinke's Diving Apparatus has also been very successfully used for obtaining phosphate of lime. This valuable substance abounds in and around many of the small islets of the South Pacific Ocean, &c., which are not inhabited, either on account of their small size or sterility. The traffic in phosphate of lime promises to become an important branch of commerce.

By means of Heinke and Davis's Diving Apparatus the guano can be obtained very cheaply and readily.

Mines. — The apparatus is also capable of an extensive application to coal, tin, lead, &c., mines, where choke damp, foul air, or water prevent the mine being worked.

Fires in mines. — Mine owners may also satisfy themselves of the exact state of unused mines, and thus obtain invaluable and reliable information as to the utility of buying them up, to pump out the water, &c., &c. In case of fire, also, in the mine, the diver can at once proceed to the course of the flame (being able to go through the smoke with perfect ease), and apply the water to the proper place.

Removal of obstructions. — Heinke and Davis also respectfully solicit the attention of Governments to their improved apparatus as a means of clearing dangerous passages from rocks, reefs, wrecks, and other obstructions. Heinke and Davis are prepared to undertake contracts for the removal of these dangers. They can command the services of the most experienced divers, and feel confident that they are in the most favourable position for carrying out submarine operations. As an instance of the practical utility of this, the late Mr. W. Heinke superintended the blowing-up and raising of the ships sunk in the harbour of Sebastopol by the Russians during the Crimean war.

Sebastopol. — The Russian Government availed itself of his services after peace was proclaimed, and the operations proved a complete success, not a vestige of the vessels remaining to impede navigation.

Robbery in Cornhill. — To show the perfection to which the apparatus has been brought, when the famous watch robbery took place at Walker's, in Cornhill, London, a woman, being hotly pursued by the police, threw a number of watches over Blackfriars Bridge into the river.

Recovery of property. — Now it would not be a very easy matter to find a watch in a soft muddy field, as the specific gravity of the gold would cause it to sink in the soil. How much more difficult, then, must it be when in the water? Nevertheless, ten of the watches were recovered from the bed of the Thames, and as some of them were of considerable value, the returns well repaid the outlay.

Heinke and Davis's apparatus is specially adapted for pearl, coral, and sponge diving, the special arrangements of the valves and pumps ensuring an equable and continuous supply of air. This at great depths is of the first importance, as the amount of work done by the diver is much lessened when he is uncomfortable and distressed by an unequal and intermittent current of air. With Heinke and Davis's apparatus the diver can work steadily and continuously, and having discovered a spot rich in coral or sponge (as the case may be) can send up the baskets [see cut] with great rapidity.

Pearl, coral, and sponge diving.

Divers equipped in Heinke & Davis's Improved Patent Diving Apparatus.
Pearl, Coral, and Sponge Diving.

RAISING SUNKEN VESSELS.

When a vessel has been sunk by collision, or otherwise, and it is found, on a survey being made by a diver, that she is not irreparably injured, she is frequently worth raising. The success of this branch of submarine engineering has lately been reduced to a certainty.

Simple method of raising vessels

There are various ways of performing this, the most simple and effective of which, when the tide has any considerable rise, is the following:—A vessel is obtained with as small a draught of water as possible, and of greater displacement than the sunken vessel. This must have but little ballast, so that great buoyancy may be the result. The vessel is taken out at low water and moored over the wreck.

The divers then descend and fasten chains or ropes to various parts of the wreck, then take the other ends of rope or chain on board and secure to crabs or blocks in vessel, in such a manner that they may easily be disengaged or taken up. These are then made fast and taut.

All cargo should be taken out of wreck to make it as light as possible, and masts sawn off.

As the tide rises, the wreck (if not too firmly imbedded in the mud) will rise also. When the tide has attained its height, the vessel or barge must be warped, sailed or steamed, as the case may be, towards the shore, until the wreck grounds again.

Nothing is now done until next low tide. The slack of chains is hauled in and the operation repeated until the wreck is above water, when it may be pumped out.

Means of increasing buoyancy.

Should there be any difficulty in obtaining a vessel large enough to float the wreck by itself, a raft may be constructed of watertight casks overlaid with planks; one of these lashed to either side of the vessel will give the required buoyancy. The size, of course, depends upon the weight to be raised. The platform thus improvised may be used as a stage for the divers to work from, and will be found very useful in cases where small vessels only can be obtained.

Should the wreck be imbedded in the sand, it may be partially eased (should there be no current) by the divers clearing away the sand, &c., from the sides. If it is found that as the tide rises the vessel does not move, the ropes or chains must be cast loose and the attempt renewed with another vessel of larger dimensions, otherwise the vessel will be overflowed by the rising tide.

Small rise of tide.

When the rise of the tide is so inconsiderable as not to be

available, it is usual to employ pontoons, which are let down full of water, fastened to the wrecks by divers and pumped out; but the following account taken from the "*Mechanic*" of the raising of the ship "Wolf" is an illustration of the employment of direct power from above. The "Wolf" was a paddle-wheel steamer belonging to Messrs. Burns, and carried the mails between Belfast and Glasgow. She was run into and sunk in seven fathoms low water by the Fleetwood steamer, "Prince Arthur."

Raising the "Wolf."

The contract for raising her was taken by Messrs. Harland and Wolf, who had successfully raised the "Earl of Dublin."

Harland and Wolf's operations.

The "Wolf" is a vessel 242 feet 7 inches long, 27 feet 2 inches broad, and 13 feet 8 inches deep; tonnage, 676; nominal horse power, 310.

It was found she had sunk in 10 feet of stiff clay, so that the rise and fall of the tide, which was only 9 feet, could not be made available.

The weight of the hull and machinery of the "Wolf" was estimated at 800 tons, and in addition to the pontoons already made for the "Earl of Dublin," two large pontoons were constructed, each having an available buoyancy of 176 tons, being $70 \times 12 \times 9 = 216$ gross. These were each divided into six compartments by watertight bulkheads, and were arranged with cocks, so that either or all could be filled with water, if required, and again pumped out. These two formed the after raft. The fore raft consisted of eight pontoons, with an available buoyancy of 500 tons. Strong logs were laid across, and fore and aft, in the manner shown in the engraving on page 233, particular attention being paid to have the distance between the tank, such as would just allow the bulwarks of the vessel to come to the position shown, and also to have an angle of the cross with the fore and aft logs exactly over each side-light. The side-lights, 25 on each side, were considered the most convenient places for fastening the lifting chains. A model side-light had been tested to 30 tons, and as each would have to bear only about 17 tons they were sufficient for the purpose. In each of the angles formed by the logs was placed a box bracket with sides of $\frac{7}{8}$ in. iron plate, carrying a hexagonal cast-iron sheave, about 10 in. diameter, fitted for the links of the chains, which were of $1\frac{1}{2}$ in. iron, and $10\frac{1}{2}$ in. long over all. This length of link was adopted to afford facility for stopping the chain by an iron cotter through the $\frac{7}{8}$ in. side plates of sheave brackets, and for shackling the chain at any link. On the top of each of the cross logs two screws were placed; these screws lay horizontally, one for each side, and were 6·0 in. length of screw, 3 in. diameter over thread, and ¾ in. pitch. The lower ends of the chains were fastened to iron hooks, of a breadth equal to one-third the circumference, and fitted to the curve of the side-lights, and these hooks, when inserted, were prevented from dropping out by plugs driven into the side-lights. These hooks were all placed early in the spring, and the chains from each ranged on the deck of the wreck, ready to be passed up without delay when the position rafts should arrive.

The wreck lay in the Lough of Belfast, about 10 miles out.

The operations connected with this plan will be better conceived from the following particulars taken from Messrs. Harland and Wolf's journal of their proceedings:—

Journal of the contractors.

July 1, Wednesday.—The fore raft of tanks left Belfast for the wreck at 3.20 p.m.; the lighter left with the logs on board. The lighter arrived first, and was moored stem and stern across and over the paddle-boxes of the "Wolf." The lighter afforded cooking and sleeping accommodation for the men. The fore-tanks, when they arrived, were brought into position, with warps to the buoys and to the lighter. Above two hundred men were then employed getting the logs out of the lighter and placed on the pontoon. They had all the logs placed on the fore raft at 10 p.m., when all hands knocked off until daylight. The after raft pontoons left Belfast at 8 p.m., and arrived at the wreck at midnight, and were secured to the S.W. anchor.

July 2, 2 a.m.—All hands turned to again to place the after pontoons, and to get the logs out of the lighter. This occupied two hours and a-half. All was now ready for the chains, and at 5.30 a.m. the divers began to send up the chains, and at 7.30 a.m. they were all up, fastened, and tightened. The tide was dead low at 8.10 a.m., and at 7.30 a.m. the men began to screw down the rafts by fifty horizontal screws. At 8 a.m. two tugs set on to tow the wreck out, and succeeded in moving her 6 ft. south, that is, broadside on to starboard, to get her out of the dock she had made. The screwing was continued while the tide kept rising, and all seemed to be going on well. As the tide turned she grounded again, and sunk down to within 8 in. of her previous draft.

At 11 a.m. a fresh breeze sprung up which raised a nasty sea. Screwing up was maintained till 5.30 p.m. Two of the lifting chains were fastened through the hawse pipes of the wreck, and at 6 p.m. these chains gave way, letting the fore ends of the two forward pontoons burst up, breaking three logs and the top of the starboard tank, whilst the port tank was bent with the extra strain which it had to bear. Water was let into the fore compartment of each of the forward tanks at once to relieve the strain on the bows, and the divers went down and recoupled the chain, passing the other end up. The tugs were set on before high water, but the "Wolf" did not move. As the tide went down the slack of the screws was taken up and more water let into the fore pontoons. Hands knocked off at 11 p.m.

July 3, Friday.—Turned to at 2 a.m.; moved one of the small after pontoons into the fore end, securing it under the logs before the foremast. Began screwing up, and high water being at 9.51 a.m., set on the tugs one hour before to pull at the stern of the wreck, and succeeded this tide in moving her her own length astern, and leaving her 5 feet higher. The morning was very calm, but it came on to blow so much that when the tide fell it was necessary to cut adrift the two fore tanks, and moor them to one of the buoys. As the tide rose in the evening it was thought unsafe to let the next fore tanks still attached have their full lift. They were half filled with water, and the fore compartments of the adjacent pontoons one-third filled, and they were left thus all night; at high water the fore pontoons were quite immersed, but that did no harm. Hands turned in at 11 p.m.

July 4, Saturday.—Hands turned to at 2 a.m., pumped out the fore tanks and tightened down the screws, making good 2 feet lift. At 8.30 a.m. the tugs were put on and moved the wreck about 500 yards, when it suddenly stopped. It was concluded that something trailing from the wreck had fouled one of the anchors. The divers went down, but could not make out what was holding. It was arranged to let the wreck and the after raft remain where they were, and to take the fore raft up to Belfast to have it repaired, and to get the two pontoons re-attached which had to be cut adrift. The screws were accordingly slacked away, and the raft taken in tow. Another tug took the two loose pontoons. The after raft was left attached to the wreck, rising and falling with the tide, lifting the stern 7 feet above the mud at high water, the bows at the same time going into the mud 9 ft. 9 in., the whole vessel vibrating on a point a little ahead of the engines.

July 5, Sunday.

July 6, Monday.—The divers went down to range the ends of the lifting chains on deck of the wreck, to be ready for the raft when it should next come down. Eight hands were at the same time sent down to better secure two of the pontoons under the logs, as they had been shifting at low water.

July 7, Tuesday.—Some hands and divers down at wreck. Divers changed two chains which had shown symptoms of weakness. Short lengths of angle irons were put under some of the sheave boxes, which had been canting over and sinking into the logs.

July 8, Wednesday.—Hands went down with the tug "Vesta," and took up the anchors lying to port and starboard of her original position, and placed them off her port and her starboard bows; also placed two light anchors amidships.

July 9, Thursday.—The fore raft having been repaired, left Belfast at 5.20 p.m., the lighter following at 6 p.m., with 70 men, and were moored at their places at the fore part of the raft until daylight.

July 10, Friday.—Hands started at 2.30 a.m., to get up chains, and shackle them to the screws. At 4 a.m. 80 more men arrived, and at once commenced to screw down the pontoons. The screwing was continued till 10 a.m., when all hands stopped for breakfast. At 11 a.m. the wreck began to rise, and tugs were put on, and the towing was kept up till 2.30 p.m., when the wreck grounded 400 yards nearer Belfast, still going stern first. At 5 p.m. commenced screwing again till 10 o'clock; made 2 feet, when all hands except the riggers turned in.

July 11, Saturday.—The tugs were put on at midnight, and towed till 3 a.m., the ship turning round, and going now bow foremost 900 yards. Began to screw at 5.20 a.m. until 10 a.m.; got in about 4 ft. of chain. Towed from 12.30 p.m. until 3.15 p.m. 1,000 yards. On beginning to screw down again it was found that the port cat head was within an inch or so of touching one of the pontoons. Divers went down to detach it, and succeeded in doing so after 3½ hours' work.

July 12, Sunday.—At 6 a.m. one of the tanks was found to be hard on the top of the after or poop deck. It was partly filled with water; taken out and moored alongside. Three of the tanks were moved a little aft under the fore raft so as to clear the projections on the main deck. These were small tanks placed under the middle of the logs. The chains were taken up 2 ft. Towed from 1 p.m. till 4 p.m. 400 yards. When the tide receded the paddle boxes were above water. At 5 p.m. men went down and tried to cut away some of the wood work from the poop deck, which was in the way of one of the tanks. Chains tightened up until this tank was hard on the deck and the logs of fore raft hard on the stern; lift 3 ft.

July 13, Monday, 1 a.m.—Towed till 4.30 a.m. 600 yards. At 5 a.m. all hands eased away the screws to take the pontoons off. Both rafts were towed off at 9 a.m., leaving one tank on the poop deck, the one the divers tried to get the wood removed from under; some of the wooden uprights of the cabin skylight had gone through its bottom. The hull was grounded on a hard bottom, and at low water the upper part of the poop deck was uncovered.

The two rafts arrived at Belfast at 11.30 a.m., the injured pontoon at 2 p.m. The rafts were brought up to get the logs raised 6ft. higher to get the vessel high enough to go into dock.

July 15, Wednesday.—Hands went down and cut off the after deck house.

July 16, Thursday.—Cleared the deck aft, took off the after winch, and removed the capstan timber heads from the forecastle.

July 18, Saturday.—Tried, unsuccessfully, to inflate the boilers with air to add to the buoyancy.

July 20, Monday.—Took a tank down fore hatch.

July 21, Tuesday.—Got the tank in its place and secured.

No work done at the wreck until

August 5, Wednesday.—Lighter started from Belfast at 3 a.m., with divers, engines, &c., to fill the boiler of the "Wolf" with air. Did not succeed in this; the divers could not make the expansion gland of the main steam pipe tight. Aft raft left the yard at 9.30 a.m. in tow of two steam tugs, getting to the wreck at 11.15 a.m. It was moored clear of the wreck until the tide would be low enough to expose the top gallant rail. The fore raft arrived at 2 p.m. It blew fresh, making it difficult to place the rafts over the wreck, but by 4 p.m. they were in place, and by 5 p.m. the chains were all attached, and all hands screwing up until 9 p.m., when they had got in about 4ft. 6in. and knocked off for the night. One of the cross logs, the fourth from the fore end, gave way, but fortunately the forecastle was uncovered, and it was shored off the deck. At 10.30 p.m., towed until 12.45 a.m., making half a mile towards Belfast.

August 6, Thursday.—Screwing down commenced at 5 a.m. After taking in 3ft. of chain, stopped for breakfast. It was not thought advisable to put more strain on the logs; levers and pumps were put under most of the logs to support them from the deck of the "Wolf," all keeping good except the one that gave way at the previous lift.

At 10.30 a.m. the tugs were set on, and towed the wreck round the Hollingwood Bank to within 500 yards of the Green Garmoyle Light, when it took the ground at 11.45 a.m., and could not be got off until the tide rose. As the tide fell the rafts took a list to starboard, and settled down about 2ft. 6in. in the mud. At 3 p.m. the rafts were screwed down until the logs were hard down; all the logs were reshored and the screws fleeted for letting go. Two of Mr. Wield's wooden pumps were got into the after stock-hole, and a gang of men started on them; at 5.30 the engine-room was empty; the water went down 9in. per hour, but the pumps had to be kept going, as it soon rose again when they stopped.

Other pumps were also put into the fore peak, and quickly pumped it dry. The tugs were put on at 10 p.m., and towed her out of her berth into deep water, and as the tide rose she went up the channel, two tugs in front and one steadying behind. The wreck was got into the Avercorn Basin at 12.30 p.m., and all hands knocked off at 1.30 p.m. on the 7th August.

The wreck — 800 tons weight — had been lifted 40ft. and towed 10 miles in 15 tides. The dock being engaged, the rafts were moored with the wreck hanging on them in the basin, until the 9th August, when it was successfully docked, and the pontoons floated off.

Cost of operations.

The expense of the operations has been £6,000.

From her long sojourn below water the "Wolf" presented a curious appearance when brought up, being entirely covered inside and out and also high up the masts with marine plants and animals, some of a most beautiful and interesting species; but with the exception of the local injury to the starboard bow caused by the collision, the hull and machinery seem to require little but cleaning, the rust on the machinery being but a very light coat, doubtless from the fact that a total submersion has a far less injurious effect on iron than that which is only partial.

In the experience of underwriters in this country, this is the first iron steamer of large tonnage that has been recovered from deep water, and the work has been watched with interest by all concerned in salvage operations.

Raising by india-rubber caissons.

Another method is to fasten large indian rubber caissons or cylinders to the sides of the vessel in a collapsed state, and when secured, inflating them with air. This alone or in combination with direct power is very effectual.

When it is desired to obtain greater lifting power, the india-rubber caissons may be filled with hydrogen gas. The buoyancy obtained by this is very great, as hydrogen is 14.5 times lighter than air. A leaden vessel is necessary to contain the zinc and sulphuric acid necessary for making the hydrogen.

Colonel Gowan at Sebastopol.

Colonel Gowan, who successfully raised many of the vessels sunk in the harbour of Sebastopol during the Crimean war, used large iron caissons 100 feet by 22 by 65, and fixed small portable engines on them; these hauled a chain passed under the wreck. He had six caissons, and the operations were very successful. He employed a great many of Heinke's diving apparatus, all of which gave the outmost satisfaction. Heinke and Davis are prepared to undertake contracts for raising sunken vessels, or to superintend works already in progress. The success of all such operations depends entirely on the diver performing his work efficiently and quickly, as all progress must be stopped when a storm arises, and if the job is in a state of transition, it will be destroyed and the labour have to be commenced afresh. Therefore a quick and experienced diver, who can take advantage of fine weather to work rapidly, is a great desideratum.

Advantage of an experienced diver.

Divers, equipped in Heinke and Davis's Improved Patent Diving Apparatus, preparing to blast Rocks and Wreck.

BLASTING.

A very important branch of submarine engineering, and one to which Heinke and Davis devote much attention, is the blasting rocks or wrecks by means of gunpowder or gun-cotton.

Best method of procedure. When such work is to be done it will always be found far cheaper in the end to obtain the services of an experienced diver, as one who is a novice at the work will cause trouble and extra expense by (1) nervousness, (2) want of experience.

The first, though not felt when above water, is very apt to be experienced when engaged on the job. This, it is unnecessary to remark, is a serious drawback. A diver who is used to the work has no such feeling, and gets his task done with greater ease and certainty.

Surveying. For those who, however, have to employ a tyro, a few hints may be of assistance. In the first place, two men should be employed to go down at once. This is easily effected, without the cost of additional apparatus, by means of Heinke and Davis's improvement for that purpose.

Small rocks. A careful survey of the rock should be taken preparatory to commencing operations. Should the rock or portion of rock to be blasted be small, a charge of powder laid at the side and heavily weighted will be found, when exploded, sufficient to destroy the obstruction, as by placing the powder thus it is enabled to explode laterally and so exert its greatest force.

Large rocks. Should the rock be large, a number of equidistant holes must be bored for the charges. The simplest mode of accomplishing this is with a long chisel and hammer, somewhat after the method of a stonemason. As the depth of the hole increases, jumping irons must be used, until a hole of the required depth is made. When this is completed another should be bored, and so on, keeping the holes in the same line, and, as nearly as possible, parallel to the face of the rock. The distance between the holes will necessarily vary with the strength of the charge, but an approximate rule may be given as follows. Distance of holes apart equal depth of holes. The depth of the holes should be from two to three feet below the level to which it is wished to reduce the rock.

Distance of holes.

Depth.

Connecting wires. When the row of holes has been completed the charges must be introduced and the battery wires connected with the fuze. The divers then come to the surface, get in the boat or barge, which must then be taken to a safe distance. When this is done (taking care not to strain or foul the battery wires) the circuit of the battery is completed and the charge explodes.

Continuing operations. When the commotion subsequent on the explosion has subsided, the divers descend again, bore fresh holes and so on till the whole rock is destroyed, blowing all projecting pieces off that have been left by former explosions. The diver, after the first explosion, should look for lateral rents and fissures in the rock in which to place the charge, as by inserting it thus a very powerful effect is obtained on explosion.

The above short summary is intended merely as an outline to show the method pursued, as it is utterly impossible in the short space allotted to this subject to give directions that would enable a diver not acquainted with the work to execute operations in a satisfactory manner. It is always the safer and cheaper plan to employ experienced divers.

MATERIAL FOR BLASTING.

Gunpowder is the usual agent employed for blasting and is fired in charges varying from 10 to 100 lbs. It is usually contained in a tin case with a fuse in the interior. The wire to connect it with the battery is brought through a cork or bung fitted into the case, which must be made water-tight by covering with a compound of melted indian-rubber, fat and resin, which must be spread carefully all over cork so as to render it completely impervious to water. The projecting wires are then connected with the battery wires and the charge tamped in. If the charge is in a vertical hole this may be most simply effected by driving in a plug of wood, taking care not to injure the wires. If a large charge is to be exploded the cask or canister should be weighted with pieces of rock, &c., seeing, however, that the case or cask is not pressed on too heavily. If this is not attended to the powder will cake and the charge not explode. Particular care always must be taken that the case is sufficiently strong to resist the pressure of the water, which at a depth of 100 feet is very considerable. *Arrangement of fuse, &c.* *Tamping.* *Weighting the powder.*

A table of pressure per square inch is given under. The pressure of the atmosphere is not included. *Pressure.*

Depth of Water.	Pressure per sq. in.	Depth of Water.	Pressure per sq. in.	Depth of Water.	Pressure per sq. in.
FT.	LBS.	FT.	LBS.	FT.	LBS.
20	9	300	134	1,100	491
40	18	350	156	1,200	536
50	22	400	178	1,300	580
60	27	450	201	1,400	624
80	36	500	223	1,500	670
90	40	600	267	1,600	714
100	44	700	312	1,700	759
150	67	800	356	1,760 or ⅓rd of a mile.	785
200	90	900	402		
250	112	1,000	446		

Gun cotton charges.

Heinke and Davis, however, recommend their friends to use in place of gunpowder compressed gun cotton charges, which have now been subjected to a long-continued and severe trial, and have proved themselves to be the safest, strongest and most economical explosive known. The principal of combining safety will force in a highly condensed form has produced invaluable results. The attention of those who are interested in Submarine Engineering is respectfully called to the following

CHARACTERISTICS OF THESE CHARGES:

Safety: Notwithstanding their enormous power they are *harmless if unconfined*, and will not explode in the open air. They are portable, convenient, and safe in use; they neither require measuring nor weighing, and do not leave dangerous fragments about. The *Pall Mall Gazette* of April 11th, 1868, says of this material,

"Not only is it quite exempt from any tendency to "spontaneous explosion, but it is also not in any way "more liable to ignition from accidental causes than "gunpowder, and it possesses the enormous recom-"mendation (not shared by gunpowder) that if fire "should reach a package of it no violent explosion can "occur, as it requires very strong confinement to "develop its explosive force."

James Wilson, Esq., Traffic Manager of the North Eastern Railway, has recently had the safety of these charges severely tested, and in his report says,

"The results of the experiments convince me that "we may safely carry Gun Cotton along with other "goods in ordinary wagons, adopting the same rules "as now apply to the conveyance of cartridges."

Stability: In any dry place the charges will retain their full strength, and may be kept in the wooden boxes in which they are packed. Should they accidently become wet they do not lose their quality, for on being re-dried they are as powerful as ever.

Strength: Bulk for bulk these compressed charges exert *six times* the power of gunpowder.

No Smoke: As these charges emit *no smoke*, and therefore necessitate no delays, the work may proceed with unusual rapidity.

Compression: The enormous force confined in a short length at the bottom of the hole allows of a much greater

	amount of work being placed before each blast than is possible with gunpowder.
Drilling:	As less Drilling is needed, there is of course a considerable saving of labour.
Health:	Whilst gunpowder is a most deleterious ingredient in the air of mines, Gun Cotton, with its freedom from smoke, bears the highest sanitary character. Dr. Angus Smith, F.R.S., in his report to Parliament says, "In every trial which the effect on the senses or the breathing, and as far as we can judge on health, was considered, Gun Cotton has come off with the highest character. I feel much confidence in speaking thus highly in its favour."
Economy:	The prices will compare favourably with the cost of gunpowder, and will be found far cheaper when the larger amount of work done in a given period is considered.

Supplied in charges of 7" 5" 4" and 2¾" diameter; any number may be placed in a hole.

The charges may be used as in powder blasting.

After the blast the air should be clear; any appearance of reek indicates that more has been put in the hole than was needed.

MEANS OF EXPLODING

Submarine Charges.

This is accomplished by the aid of electricity. There are various contrivances for doing this, the most common of which is the galvanic battery. There are various forms of these, the principal of which are, Groves', Bunsen's and Smee's.

Groves' is a zinco-platinum battery, Bunsen's a zinco-carbon, and Smee's a zinco-platinum. Of these, Groves' and Bunsen's are the most powerful, but all are liable to one great objection when used at sea, viz., the indispensable employment of acids, which when the boat or barge is tossing about are apt to spill. They also require to be kept clean, and have to be prepared afresh every time they are used. As they, however, are still occasionally employed, Heinke and Davis supply them.

Heinke and Davis, however, recommend to their customers a much more portable and effective arrangement in their magneto-electric exploder, as represented in woodcut below.

Heinke and Davis's magneto exploder.

It consists externally of a mahogany box, screwed down on a stand. The only parts visible outside are the button and keeper.

It is kept covered, so that dust, water, &c., may not injure it. Internally, the arrangement consists of a large electro-magnet; at either pole there is a coil of copper wire which acts as an induction coil. The keeper locks the machine and prevents the circuit from being broken. The wires connected with the submarine charge are secured with the binding screw. The keeper is drawn out, and the button pressed down for an instant. The circuit is interrupted and the current made to pass down the wires and the charge explodes. Heinke and Davis supply these usually to fire 8 fuses in a circuit, but they may be made any size. The advantages claimed for this magneto-electro exploder over the ordinary battery are the following:—

1s. It is always ready.
2nd. Requires no preparation.
3rd. Cannot get out of order.
4th. More portable.
5th. More sure in its action.

Abel's fuses.

It is used in conjunction with Abel's fuses, which are too well-known to require any explanation. They are more certain in their action than any other fuses, and are specially adapted for submarine operations. Heinke and Davis supply the fuses, insulated wire, and all requisites for submarine blasting.

INSTRUCTIONS FOR DIVING.

Heinke and Davis append a few practical instructions for using their improved apparatus, which will be found of assistance to those who are not acquainted with its working, although it is not at all advisable to be guided entirely by them. The safe plan is to make a few descents under the instruction of an experienced diver.

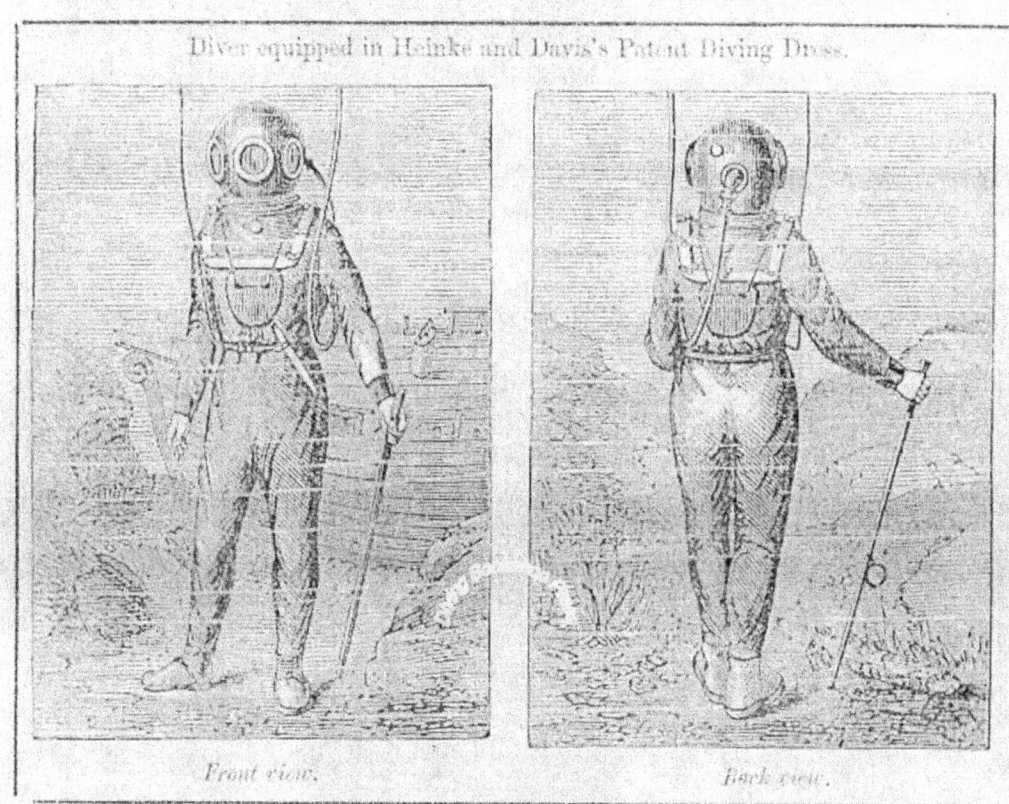

Diver equipped in Heinke and Davis's Patent Diving Dress.
Front view.   *Back view.*

AIR PUMP.

1. The Air Pump chest must rest on a flat surface to prevent oscillation. If worked at sea it should be lashed to supports by means of ropes passed through the iron rings.

2. All parts of the pump to be kept clean and bright, and lubricated very slightly with sweet oil. Bad or coarse oil not to be used, as the smell is very unpleasant to the diver.

3. Fix on handles and fly-wheel, and unscrew caps in front and back of pump. Turn the handles a few times to see that all parts work easily.

4. Connect hose to pump.

Helmet.

5. To be carefully examined to see that it is in order. Back and front valves to work easily.

Dress.

6. To be carefully examined and washed in fresh water each time after using.

7. To be on *no account* dried before the fire or in the sun.

8. It must not be oiled or greased.

9. For repairing a tear in dress proceed as follows: See that part to be repaired is quite dry, then cut a piece of prepared canvass of sufficient size amply to cover the damaged place. Spread a thin coat of the solution on the dress, when this is dry spread another over it, and before this last is quite dry put on the prepared canvas and press it well down. Smooth it all over to ensure perfect adhesion, and place weights on it till dry.

Hose.

10. To be kept coiled in a cool dry place.

11. The unions should be wrapped in pieces of rag to keep them from being damaged. When in use one end of hose is screwed on pump connection and the other on helmet.

12. See that all unions are well screwed up.

13. If hose is damaged, cut out injured part, and extract from each end the spiral wire end and cut off about two inches. Then insert one of spare unions in hose, and bind round with fine wire.

Note.—A set of spare unions and hank of wire sent if required.

Dressing Diver.

14. First put on jersey, cap, drawers and stockings, then get on waterproof dress as far as the waist.

15. The diver must then hold up his arms and the attendant pull up dress.

16. The diver then puts in first one arm and then the other.

17. Over this, duck jacket and overalls. Then boots.

18. If the cuffs are too large they must be tied with bands or rings.

19. Take care not to pass hands too roughly through the vulcanized cuffs so as not to tear them.

20. The breast plate of helmet then to be fastened to dress by taking off helmet bands, and passing studs through the holes in the vulcanized india rubber bib of dress one by one, then place the four helmet bands over studs, seeing that they are arranged as numbered, and screw on fly nuts.

21. Top of helmet (with hose connected) then to be screwed on, without front glass.

22. Pumps to commence working.

23. Weights to be put on back and front and fastened with a slip knot.

24. Signal line tied round waist

25. Screw on eye glass, first asking if all is right.

26. The diver is now ready to descend. This is usually done by means of a ladder, which must be fastened firmly to the barge or landing stage from which the diver works.

27. The signal man must hold the life line in one hand and guide the descent of the hose with the other, care being taken not to let the hose or life line get too slack.

28. The diver should take down with him a coil of small line to fasten to bottom of ladder to be used as a clue to guide him back.

29. It is indispensable that the signal man be steady and intelligent, as the diver's life might be lost through his negligence. He must not be spoken to whilst the diver is beneath water. From time to time he should give one distinct pull at the life line. This must be answered by the diver by another distinct pull. **Should the diver not answer he is to be pulled up immediately.**

The pumps must be kept working uninterruptedly from the time the diver's helmet is put on until it is taken off. **This is most important, and must be attended to by the signalman.**

Management of Front Valve.

Before descending let the front valve be about half shut.

If more air is wanted close the valve a little.

If less open the valve a little.

If there is a difficulty in keeping the feet on the ground, open the valve.

Should the diver wish to rise without assistance, close the valve entirely.

If the glass should break or dress tear, close valve and signal "pull up" at once.

The back valve in Heinke and Davis's helmet is provided with a set screw, so that it can be opened or shut at pleasure.

When diver is finished, hose and dress should be dried as directed, the helmet cleaned and packed away in box until again required.

The bright parts of pump to be wiped, handles and fly-wheel removed, and pump box locked.

The duck overalls also should be dried.

HEINKE & DAVIS'S
IMPROVED PATENT DIVING APPARATUS

Consists of the following:

Heinke and Davis's improved three-throw air pump and fire engine combined, with gun-metal cylinders and valves, wrought iron crank and handles. Bourdon's Patent Pressure Gauge cast iron fly wheel, fitted in teak chest, with iron rings for lashing.

Heinke and Davis's improved tinned copper helmet, with back and front valves, brass eye frames and plate glass eye glasses, brass bands, wing nuts, connexions, &c.

Two patent double tanned twilled waterproof vulcanised india-rubber diving dresses, with vulcanised cuffs and bib. Best make.

100 feet of vulcanised extra strong india-rubber hose, with interior spiral wire fitted with patent double capped safety unions.

100 feet of signal line.

Complete suits of under clothing, consisting of guernseys, drawers, stockings, cap, and pad.

Two lead weights, rigged with ropes, and brass thimbles to fasten on helmet.

A pair of leather diver's boots with lead soles.

Duck overalls consisting of jacket and trowsers to go over waterproof dress.

A knife, belt and sheath. Best make.

Tool chest containing spanners, screw driver, oil can, washers, punch, &c.

Price complete as above.

Extras.

Heinke and Davis's Patent Arrangement for sending down two divers at once with one pump.

Two 20 feet lengths of Improved Vulcanised India Rubber Tubing; large size for fire engine.

Unions for ditto

Gun Metal Nozzle for ditto, best make.

Gun Metal Rose for ditto ,,

Heinke and Davis recommend their friends, especially those who intend to use the apparatus in distant parts, to order a few spare nuts, washers, springs, helmet glasses, &c., parts which are liable to be lost or mislaid, the want of which cause much inconvenience.

Improvements and Additions Made By
HEINKE AND DAVIS
IN THEIR "PATENT DIVING APPARATUS."

Air Pump. — An arrangement by which the Air Pump may be used as a fire engine. This is designed specially for the use of ships. It can be used equally as an air pump, fire engine, or auxiliary pump to pump water to boilers should the engine pumps get out of order, to swill the decks and to wet the sails, and will be found to overcome the objections of captains to the ordinary air pump, viz., that it was seldom used. Heinke & Davis make no extra charge for this patent improvement, so that over and above the reduction in price, they supply a more useful machine than any other makers.

Duplicate Improvement. — By a simple and effective improvement, two men can be sent down at once from one pump when working at the same depth. The saving effected by this is very great, as by it the services of two pumpers at £1 per week each are dispensed with; the saving therefore is £100 per annum and cost of extra pump.

Helmet Valves. — Heinke's Patent Front Valve is retained. In the back valve Heinke and Davis have made an improvement by which it can be closed, or opened at the option of the diver whilst under water.

Eye Frame. — An improvement by which all risk of dropping the eye frame in the water is obviated by means of a small chain and gallery.

Various other important improvements have been perfected, which will be duly announced and described as soon as the necessary steps have been taken to protect them.

HEINKE AND DAVIS SUPPLY
COMPLETE DIVING APPARATUS AS ABOVE.

Diving Bells,
Cranes,
Crabs,
Rope Ladders,
Tools for Cleaning Ships' Bottoms,
Barges,
Voltaic Batteries,
Magneto-Electric Exploders,
Abel's Fuses,
Gunpowder,
Gun Cotton,
Electric Lights,
Insulated Wires,
Boring Tools,
All sizes of Vulcanized Hose,

India rubber caissons or bags for raising sunken Vessels, and every description of goods necessary for Submarine works.

Heinke and Davis undertake all descriptions of Submarine work.

Diving Apparatus Lent on Hire.
Men instructed in the Art of Diving.

Heinke and Davis beg to call the attention of Ship Owners, Engineers, Merchants, and Contractors to the use of their Apparatus for the following purposes:—

Repairing Ships when at Sea,
Recovering Property from Wrecks,
Making Foundations of Bridges, Piers, Docks, Breakwaters, Harbours, and Sea Walls, and for repairing and inspecting the same.
Pearl, Coral, and Sponge Diving.
Surveying Mines.
Repairing Artesian Wells.
Blasting Rocks, Reefs, &c.
Blowing up Wrecks.
Repairing Gas and Waterworks.
Obtaining Guano.
Repairing and Erecting Lock Gates, and all description of Submarine Harbour Work.
Recovering Sunken Treasure.
Surveying and Repairing Canal Banks and Walls, and for all Submarine Operations, or Works in which foul air, gas, or liquid prevent ordinary labour.

In conclusion, HEINKE and DAVIS beg the favour of an inspection of their Improved Diving Apparatus, at their Works, **2, Brabant Court, Philpot Lane**, where every information will be given.

Waterlow & Sons, Printers, Great Winchester Street, E.C.

My Diving-Dress

By One Who Has Done With It

This entertaining little piece first appeared in an early edition of the Strand Magazine. It has been included here as it is a wonderful example of the popularity of the theme of 'deep-sea diving' as presented to the general public of that time. The article is made doubly enjoyable by the inclusion of the amusing illustrations.

Another reason for its inclusion is for its historic relevance in that it refers strongly to Whitstable in England for this is considered to be the birth place of the Hard Hat diving industry. This is the place where the new invention of the diving dress was first embraced wholeheartedly by much of the population of the harbour town. So much so that it resulted in the expansion of diving operations throughout the United Kingdom and later to the rest of the world.

Reading through it strongly suggests that it is a fictional piece presented as fact. However, the average reader of the day would have most likely believed it to be an amusing and true account of a diving exploit.

It is important to remember that in the 1800's helmet divers were at the cutting edge of technology. The men who went down to wrecks and carried out a huge variety of salvage work did so at great risk to their own lives. Underwater salvage work is a complicated affair even today despite the huge advances in technology. Imagine then, the incredibly difficult tasks that these early divers performed.

A.J.D

My Diving-Dress.

By One Who Has Done With It.

A LARGE part of my life has been spent in seeking and experiencing novel sensations. Precisely what quality of mind it is that urges me to try experiments with myself and other things I do not know positively; but I firmly believe it to be dauntless intrepidity. My fond mother, in early days, used to call it a noble thirst for information, and predicted for me a life of scientific eminence; other people have been so ill-natured as to call it abject imbecility, and to predict an early grave from a broken neck or a dynamite explosion, or something equally sensational and decided. Never mind what it is. In boyhood's days it led me once up the chimney, once on a river in a wash-tub, once down a gravel-pit with a broken head, and frequently across my father's knee, with a pain in another place. Since I have arrived at years of discretion (or greater indiscretion —just as you please), it has taken me up in a balloon, out to sea in a torpedo-boat, up the Matterhorn (with no guide but a very general map of Europe, having the height of the mountain marked on it in very plain figures), along Cheapside on a bicycle at mid-day, to a football match in the capacity of referee, and lastly, and most recently, down under water in a diving-dress. Many of these experiences were sharp enough while they lasted, and the diving was as disturbing as most; but, still, I believe nothing was quite so uncomfortable as the football refereeship.

But, just now, I am concerned only with the diving. I have been now and again to Whitstable, where, I believe by some remarkable process of Nature, every third male person is born a diver. Anyway, Whitstable is the place where divers mostly grow, and where I caught the temptation to go a-diving myself. I should feel grateful to any obliging Anarchist who would blow up Whitstable tomorrow.

I mentioned my desire to one or two old divers who had permitted me to make their acquaintance in consideration of a suitable succession of drinks, but met with jeers and suspicion. I believe they were afraid of opposition in the business. But Whitstable never produced a diver that could put me off. I took the royal road. I bought a diving-dress for myself—how much I paid I shall not say here, for why should an unsympathetic world measure my lunacy by pounds, shillings, and pence?—especially as that would make rather a long measurement of it. Never mind what I paid. I got the dress, and I also got permission to go down and amuse myself on a sunken coasting vessel lying off Shoeburyness.

"MY DIVING-DRESS."

It was a very noble diving-suit, and the new india-rubber squeaked musically as I moved, and smelt very refreshing. There was a shield-shaped plate, rather like a label on a decanter, hanging on my chest, that would have looked more complete with "Whisky," or some similar inscription, on it. There was a noble metal collar—about thirty-two, the size would have been, on the usual scale. I had also a very fetching red night-cap, while my helmet was a terror to all beholders. I don't mind confessing to a certain amount of discomfort while they were building me up in this dress—partly due to a vivid imagination. The helmet made me think of the people in the story who put hot-pots on the heads of strangers, and I seemed stifling at once. What if I were unpacked at last from this smelly integument—a corpse? But this was unmanly and undiverlike. There wasn't much comfort to be got out of the leaden shoes—try a pair for yourself and see—but when all was ready I made a shift to get overboard and down the ladder provided. It was not a great deal of the outer world that I could see through my windows, and I hung on to that ladder with something of a desperate clutch. When at last the water stretched away level around my windows, then, I confess, I hesitated for a moment. But I made the next step with a certain involuntary blink, and I was under water. All the heaviness—or most of it—had gone out of my feet, and all my movements partook of a curiously easy yet slowish character. It looked rather dark below me, and I tried to remember the specific gravity of the human body in figures by way of keeping jolly. At the top of my helmet the air-escape-valve bubbled genially, and I tried to think of myself as rather a fine figure of a monster among the fish, with a plume of bubbles waving over my head. You do think of trivial things on certain cheerful occasions. Remember Fagin in the dock, for instance.

"I WAS ENGULFED IN AN AWFUL CONVULSION."

It was not as long as it seemed before I was on the wreck, and down below in the nearest hold. Regular professionals had already been at work, and access to different parts of the ship had been made easy. Now, in this big hold was an immense number of barrels, stood on end and packed tightly together—barrels of oil, to judge from externals. I tried to move one, but plainly they were all jammed tightly together, and not one would shift. I took the light axe with which I had furnished myself, using it alternately as wedge and lever, and at last felt the barrel move. I had certainly loosened it, and pulled up the axe with the intention of trying to lift the barrel, when I was suddenly engulfed in an awful convulsion as of many earthquakes in a free fight. The world was a mob of bouncing oil-barrels, which hit me everywhere as I floundered in intricate somersaults, and finally found myself staggering at the bottom of the hold, and staring at the roof, whereunto all the barrels were sticking like balloons, absolutely blocking up the hatchway above me.

"IN THE HOLD WERE AN IMMENSE NUMBER OF BARRELS."

What was this? Some demoniac practical joke of fiends inhabiting this awful green sea about me? Were they grinning at me from corners of the hold? or had some vast revolution in the ways of Nature taken place in a second, and the law of gravity been reversed? It was not at all warm down there, but I perspired violently. Then a notion flashed upon me. Those barrels must have been *empty*. Jammed together, they stayed below, of course, but once the jam was loosened they would fly at once towards the surface. Then I thought more. I had been an ass. Of course, those barrels would do as they had done, even were they full of oil. Oil floats on water, as anybody should know. They might be either full or empty, it didn't matter a bit. I had forgotten that I was moving in a different element from the air I was used to, where barrels of oil did *not* incontinently fly up into space without warning. Obviously, I had made a fool of myself, but I had some comfort in the reflection that there was nobody about to see it. Then it came upon me suddenly that I would rather have someone there after all, for I was helpless! Those horrible barrels were having another jam in the hatchway now, and my retreat was cut off entirely. Here I was like a rat in a cage, boxed in on every side. My communication-cord and my air-pipe led up between the barrels, to outer safety; but what of that? I perspired again. What would happen to me now? Why did I ever make a submarine Guy Fawkes of myself, and thus go fooling about, where I had no business, at the end of a flexible gas-pipe? If I could have dated myself back an hour at that moment, I believe I should have changed my mind about going in for this amusement. At this, I began thinking about trivial things again—how, paraphrasing a certain definition of angling, diving might be described as matter of a pipe with a pump at one end and something rather worse than a fool at the other. I determined, if ever I got out alive, to fire off that epigram at the earliest possible moment — so here it is.

I made an effort, pulled myself together, and determined on heroic measures. My axe lay near, and, with a little groping, I found it. I would hew my way out of this difficulty through the side of the vessel. I turned on the inoffensive timbers at my side and hacked away viciously—with, I really fancy, a certain touch of that wild, stern, unholy joy that anyone feels who is smashing somebody else's property with no prospect of having to pay for it. Every boy with a catapult, who lives near an empty house, will understand the feeling I mean—especially if the empty house has a large conservatory.

The timbers were certainly stout. The work was a bit curious to the senses—the axe feeling to work with a deal more dash and go than the arm that directed it. At any rate, the exercise was pretty hard. Any millionaire in want of an excellent, healthy, and expensive exercise should try chopping his way through the sides of ships—it will do him a world of good, and will be as expensive as anybody could possibly desire. After a while I found I had well started a plank, and, once through, chopping away round the hole was not so difficult. Still, when I had a hole big enough to get through, I did not feel by any means as fresh as I had done when first that horrible copper pot was screwed down over my head.

I squeezed through the hole, and at the first step I had ever made on the real sea-bottom, I fell a savage and complicated cropper over my communication-cord. I got up, but, as I stepped clear of the cord, a frightful conviction seized my mind that I was a bigger fool than I had ever given myself credit for being. What in the world was the good of getting out through the side of the vessel when that communication-cord—my only means of signalling—and that air-pipe — my only means of submarine life—led up through the boat itself and among those execrated oil-barrels? Awful! Awful! I sat down helplessly on a broken rock and stared blankly through my windows. To weep

"I HACKED AWAY VICIOUSLY."

"AWFUL!"

would have been mere bravado, with so much salt water already about me. I tried to signal with the communication-cord, but it was caught somewhere in that congregation of oil-barrels. It seemed to be all up, except myself, who was all down, with no prospect of ever rising in the world again. Shadowy forms came and went in the water about me, and I speculated desperately in how long or how short a time these sea-creatures would be having a dinner-party, with *me* as the chief attraction. I wondered, casually, whether the india-rubber would agree with them, and hoped that it would not. Then I wondered what they would take for the indigestion, and I thought they would probably take each other—it's their way, I believe. I was wandering on in this way, and had just feebly recollected that there was four pounds eight and something in my pockets above, which was a pity, because I might have spent it first, and that I owed my landlady fifteen-and-six, which was a good job, because it would compensate for that claret she said the cat drank, when an inspiration seized me—a great inspiration. I should probably have called out "Eureka!" as did the venerable discoverer of that principle of specific gravity that had lately (literally) taken a rise out of me, if I had thought of it, but I didn't, which was fortunate, because it is rather a chestnut after all.

This was my notion—a desperate one, but still one with hope in it. I would shut off the air-escape valve on my helmet, so that the air being pumped in would inflate my india-rubber dress like a bladder. Then I would cut my air-pipe and communication cord, stuffing the pipe and tying it as best I might, take off my leaden shoes and rise to the surface triumphantly, like an air-cushion, or, say, an oil barrel. Specific gravity having taken a rise—all the rise—out of me, I would proceed to take a rise out of specific gravity; a great, glorious, and effective rise to the upper world. No office-boy on promotion ever looked forward to his rise with more hope than I to mine. It was a desperate expedient certainly, but what else to do?

I took off one leaden shoe and loosened the other, ready to kick away. I shut the escape-valve. I cut the cord with my axe on the rock I had been sitting on, and then, when the air had blown out my dress to most corpulent proportions, I took the decisive stroke. I chopped through the air-pipe. I stuffed it as well as possible and tied it in some sort of a knot—it was *very* stiff—in a great hurry, and then—I kicked off the leaden shoe.

"I TOOK OFF ONE LEADEN SHOE."

Never, never, never—even if I live on Jupiter after this planet is blown to shivers—shall I forget the result of my forlorn-hope dodge. I kicked off the shoe, as I have said, and, in an instant, the whole universe of waters turned upside down and swirled away beyond my head. In sober fact, *I* had turned upside down—as I might have known

"I HAD TURNED UPSIDE DOWN."

I should do, if only I hadn't been a bigger fool than ever.

Of course, the moment my leaden shoes went, *down* came my copper head-pot, being my heaviest part, and up went my feet. I had a pretty quick rise, certainly, but I prefer not to recall my feelings during the rush. I can quite understand now why a rise in the world makes some people giddy. All that I had before felt of amazement and horror, I now felt multiplied by fifty and squeezed into about two seconds, so that they felt like ten hours. Up through that awful water and those moving shadows I went, feeling that I was in reality held still, like a man in a nightmare. When at last I stopped, I felt that it was but a matter of moments, and the air would leak away through that cut tube, and I should go down again, still head under, for the last time, to die in that grisly combination of mackintosh and copper kettle; also I felt choking, stifling, when—something had me roughly by the ankle, and I was dragged, a wretched rag of misplaced ambition, into a boat. The appearance of my legs sticking out above water had, it seemed, caused intense amusement among the boat's crew—a circumstance which probably ought to have gratified me, although it didn't.

I have little more to add, except that I shudder, to this day, whenever I see an acrobat standing on his head, because it is so graphically remindful. But, if anybody is thinking of going in for diving by way of placid enjoyment, I shall be delighted to treat with him for the sale and purchase of a most desirable diving-dress in unsoiled condition, cut in the most fashionable style, with a fascinating copper helmet and commodious collar, and a neat label for the chest. The shoes will not be included in the bargain, having been inadvertently left in a damp place.

© The Diving Bookshop Press, UK 2006